Table of Contents

Getting Started

The next time you want to create a special centerpiece or unique homemade gift, just whip up an easy-to-make fruit bouquet that is both incredible and edible!

Practice Food Safety

Wash your hands thoroughly and often with soap and water when handling fresh fruits and vegetables. In addition, make sure your work surface is clean and sanitary.

Under running water, rub fruits and vegetables briskly with your hands to remove dirt and surface microorganisms. Prior to cutting or peeling fruits and vegetables such as melons, carrots or pineapples, scrub the outer hard rind or firm skin under running water with a vegetable brush.

Waxes are often applied to certain produce, such as apples, cucumbers and zucchini, to help retain moisture. Therefore, do not wash these fruits and vegetables until you are ready to create your fruit bouquets in order to keep them firm and crisp as long as possible. After washing, pat the produce dry with paper towels to remove any moisture.

It is important to keep most fruits and vegetables at a cool temperature while preparing and arranging your bouquet. After pieces of the bouquet are cut, refrigerate them as directed until ready to assemble the bouquet. For optimal freshness and beauty, it is highly recommended that your bouquet be displayed and served the same day it is prepared. If you intend to display your bouquet as a centerpiece, assemble and serve it as close to the beginning of the event as possible. If your gift bouquet needs to be transported, cover it loosely with a large food-safe plastic bag and pack it securely in a large cooler. Encourage the recipient to enjoy the produce as quickly as possible, and to store any leftovers tightly covered in the refrigerator.

Prepare the Base

Part of the fun of creating your bouquet is choosing an appropriate, colorful container to hold the fruit flowers. Choose containers that are at least 5″ tall with a sturdy, flat base and wide-mouth opening. When available, use a container that has a plastic liner. Wash and thoroughly dry any container and liner before filling it with fruit flowers.

A head of iceberg lettuce is a great, inexpensive way to secure the fruit flowers in your bouquet. It can be easily cut to size to fit most containers. For very large containers, it may be appropriate to use more than one head of lettuce. For very small containers, the lettuce can be torn into large pieces and layered in the container. It is not recommended to use cabbage as a base since the surface is hard to puncture.

Choose Flower Stems

A number of items can be used as flower stems. Thin plastic sticks, long drink stirrers and bamboo or plastic skewers would all work as appropriate flower stems. However, 10-inch bamboo skewers, which were used to make all the arrangements in this book, are the recommended material to use as flower stems. They are inexpensive, widely-available and easy to cut down to size, if necessary.

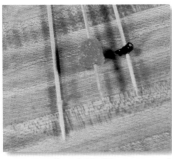

The longer the fruit flowers sit in the arranged bouquet, the more likely they are to begin sliding down the stems. This is especially true for heavier flowers, such as the pineapple or melon daisies shown on pages 6 and 18. To prevent the flowers from sliding down the skewers, other items can be slid onto the skewer before the flower is placed. Wrap small craft rubber bands approximately 2″ from the pointed end of a skewer to make a ridge. If you prefer to use a food item, slide a raisin or gumdrop onto a skewer to act as a stopper.

You will need:

- 2 to 3 whole fresh pineapples
- 1 small cantaloupe
- 1 small bunch green grapes
- Flower-shaped metal cookie cutters
- Melon baller
- 20 to 25 (10″) bamboo skewers
- 1 head iceberg lettuce
- 1 small bunch purple or green kale or leafy lettuce
- 1 medium oval basket

Fruit
Bouquets

Create Your Own Gifts & Centerpieces

Delicious Designs

Printed in China
by G&R Publishing Co.

Published By:

507 Industrial Street
Waverly, IA 50677

ISBN-13: 978-1-56383-298-7
ISBN-10: 1-56383-298-4
Item #3621

To Begin...

1 Begin by slicing a pineapple sideways into ¾"- to 1"-disks. Cut the pineapple over a sheet pan with a rimmed edge to catch the juice. For a medium-size bouquet, you will need 13 to 15 pineapple disks. To cut the flowers, center one of the flower-shaped cookie cutters over a pineapple disk. (Metal cookie cutters are recommended for a clean, even cut.) Press straight down on the cookie cutter, using even pressure.

2 Turn the pineapple disk over and gently press the flower shape out of the disk. Cut the remaining disks into flowers using various sizes of flower-shaped cookie cutters. Place the pineapple flowers in an even layer on a clean rimmed baking sheet; place in the refrigerator to chill while assembling the remaining pieces. Any remaining pieces of the cut pineapple disks can be discarded.

3 To make the flower centers, cut the cantaloupe in half and remove the seeds. Using a melon baller, cut balls from the orange cantaloupe flesh. The balls can be either completely round or they can have one slightly flat side. Cut enough balls to have one for the center of each pineapple flower. Place the cantaloupe balls on a plate and refrigerate to chill while assembling the grape spears.

4 To make the grape spears, thread 4 or 5 similar-size grapes onto a wooden skewer, starting at the stem-side of each grape and piercing straight through to the bottom end of each grape. Do not pierce all the way through the final grape on each spear, allowing the skewer to remain concealed. For a medium-size bouquet, you will need 6 to 8 grape spears. Place the grape spears on a plate and refrigerate to chill while assembling the base.

5 While the fruit pieces are chilling, prepare the lettuce base. Cut the head of lettuce as necessary to fit easily into the basket. For a medium oval-shaped basket, such as that used in the photo, cut about 1½" from both sides of the lettuce head. If necessary, use the cut-off pieces to fill the bottom of the basket. Place the lettuce head in the basket so the top of the lettuce sits 1" to 2" above the rim of the basket. Stick purple or green kale leaves into the basket around the lettuce. Continue adding kale until the lettuce is completely covered. The skewers of fruit added later will hold the kale in place.

6 To assemble the daisies, remove the pineapple flowers and cantaloupe centers from the refrigerator. Pierce the hard center of one pineapple flower with a skewer, pressing from the bottom side of the flower, through the center, to the top side of the flower, allowing about ½" of the skewer to be exposed on the top side. Press one cantaloupe ball onto the exposed end of the skewer, stopping before the skewer pierces the top end of the ball. If necessary, use rubber bands, gumdrops or raisins to keep the flowers from sliding down the skewers, as explained on page 5.

7 Arrange the daisies in the basket, starting with the bigger blooms around the bottom and filling in with the smaller blooms as the arrangement fills up. Pierce the skewers through the kale and into the lettuce until the flowers sit at desired height. If necessary, cut the skewers down to an appropriate height using nail clippers.

8 As the basket is filling with daisies, carefully stick the grape spears into the arrangement to fill any gaps. When you are happy with your bouquet, carefully package it for delivery or return the entire arrangement to the refrigerator until ready to display and serve.

Try these variations:

- Make flower centers out of honeydew rather than cantaloupe, or use blueberries to create the center of each flower.
- Use purple grapes instead of green grapes for a darker contrast.
- Create a two-layer affect by placing a small pineapple flower over a large pineapple flower on the same skewer.
- Use a basket with a handle, arranging the pineapple flowers so the handle stands up when the arrangement is on display.

Sweet Kisses

You will need:

- 1 whole pineapple
- 25 to 30 strawberries
- 8 to 10 caramel-filled chocolate kisses
- Heart-shaped metal cookie cutters
- 30 to 40 (10″) bamboo skewers
- 1 head iceberg lettuce
- 1 small bunch purple or green kale or leafy lettuce
- 1 large vase, ceramic planter or urn

To Begin...

1 Begin by slicing a pineapple sideways into ¾"- to 1"-disks. Cut the pineapple over a sheet pan with a rimmed edge to catch the juice. For a medium-size bouquet, you will need 6 to 8 pineapple disks. To cut the hearts, center one of the heart-shaped cookie cutters over a pineapple disk. (Metal cookie cutters are recommended for a clean, even cut.) Press straight down on the cookie cutter, using even pressure.

2 Turn the pineapple disk over and gently press the heart shape out of the disk. Cut the remaining disks into hearts using various sizes of heart-shaped cookie cutters. Place the pineapple hearts in an even layer on a clean rimmed baking sheet; place in the refrigerator to chill while assembling the remaining pieces. Any remaining pieces of the cut pineapple disks can be discarded.

3 Next, create the strawberry buds. Rinse the strawberries under cool water and pat dry gently with paper towels. Choose large, full strawberries of a similar size. If desired, remove the stem and leaves from each strawberry, however it is not necessary to do so. Poke the pointed end of a skewer into the stem end of a strawberry, stopping before the skewer pierces through the berry; repeat with remaining berries and skewers, then refrigerate. If desired, skewers can hold two, three or even four strawberries for added height.

4 For a large, tall vase, such as that used in the photo, it is necessary to fill up the bottom of the vase with material before inserting the lettuce base. Place tightly-packed folded kitchen towels in the vase until it is a little more than halfway full. If you do not want to use towels, the base can be filled with cut-to-size Styrofoam or additional lettuce pieces.

5 Once the vase is more than halfway full, cut the lettuce head to fit in the vase. Place the lettuce head in the vase so the top of the lettuce sits about 1″ above the rim of the vase. Stick purple or green kale leaves around and over top of the lettuce to cover. The skewers of fruit added later will hold the kale in place.

6 To assemble the bouquet, begin by inserting a skewer into the pointed side of each pineapple heart. Slide the heart vertically down onto the skewer until the skewer hits the hard center of the pineapple piece. Next, place the hearts in the bouquet. Do this by piercing the bottom end of the skewer through the kale and into the lettuce base. Place the larger hearts near the center and the smaller hearts on either side.

7 Once all the pineapple hearts are in place, stick the strawberry buds into the bouquet to fill any gaps. Place the buds close together, in order to cover as much of the kale as possible. Turn the vase often while arranging to make sure it is filled evenly on all sides. Sometimes, it is easier to place the strawberry buds in the bouquet by first sticking a skewer into the kale and then sliding the strawberry onto the skewer.

8 Add the final touch by garnishing the arrangement with kiss blossoms. Carefully stick the pointed end of a skewer into the bottom flat side of one chocolate kiss; repeat with the remaining kisses and skewers. Stick the kiss blossoms into the small spaces between the pineapple hearts and strawberry buds.

Try these variations:

- Instead of caramel-filled chocolate kisses, use peanut butter-filled, coconut-crème or cherry cordial chocolate kisses. Any small wrapped candy with a soft filling will work.
- This bouquet makes a unique and delightful Valentine's gift for a special recipient. Accompany the bouquet with a note that says, "Showering you with sweet kisses!"
- Make this bouquet even sweeter by dipping half of each pineapple heart in chocolate. Allow the chocolate to cool and harden before inserting the skewers.

Dipp'n'Dots

You will need:

- 2 Clementines
- 2 bananas
- 1 pint blueberries
- 1 pint raspberries
- 1 small bunch green and/or purple grapes
- 20 to 25 (10˝) bamboo skewers
- 1 head iceberg lettuce
- 1 small to medium tall canister

To Begin...

1 Begin by making the grape spears. Thread 5 or 6 similar-size grapes onto a skewer, starting at the stem-side of each grape and piercing straight through to the bottom end of each grape. Do not pierce all the way through the final grape on each spear, allowing the skewer to remain concealed. For a small-size bouquet, you will need about 10 grape spears. Place the grape spears on a plate and refrigerate to chill while assembling the remaining spears.

2 To make the blueberry spears, thread 8 to 10 similar-size blueberries onto a skewer, starting at the bottom of each berry and piercing straight through to the top end. Do not pierce all the way through the final berry on each spear. For a small-size bouquet, you will need 4 to 6 blueberry spears. Create about 5 raspberry spears in the same fashion, threading the berries upside down onto the skewers. Place the berry spears on a plate and refrigerate to chill.

3 Peel the Clementines and divide the fruit into sections, removing any white membrane or pith from the sections. To make the Clementine spears, thread 4 or 5 sections vertically onto a skewer, alternating the direction each section faces. For a small-size bouquet, you will need 3 to 5 Clementine spears. Place the spears on a plate and refrigerate to chill.

4 Peel the bananas and cut them into ½″ rounds. Thread 4 or 5 rounds onto a skewer, starting at the side of each round and piercing straight through to the other side. Do not pierce all the way through the final banana round on each spear. For a small-size bouquet, you will need 3 to 5 banana spears. Set aside the banana spears until ready to assemble the bouquet.

5 Next, prepare the lettuce base. Cut the head of lettuce into round thick pieces and layer them into the canister. Fill the canister ¾ full with tightly-packed lettuce. Do not extend the lettuce above the rim since the spears will be stuck down into the canister. If desired, cover the lettuce pieces with green or purple kale.

6 Remove all the spears from the refrigerator and arrange them in the canister, alternating different fruit types. Turn the canister often while arranging to make sure it is filled evenly on all sides. Place the shorter skewers around the edge with the taller skewers in the middle. When you are happy with your bouquet, carefully package it for delivery or return the entire arrangement to the refrigerator until ready to display and serve.

Chocolate Dipping Sauce

Add the final touch to your Dipp'n Dots bouquet by presenting it with this fondue-style chocolate sauce. It is so easy to make and will stay at a good dipping consistency for about 45 minutes. To re-melt the sauce, simply microwave it for 20 seconds and stir.

Ingredients

½ C. butter
1 (14 oz.) can sweetened
   condensed milk
6 to 8 oz. chocolate chips

Directions

In a medium saucepan over low heat, combine the butter, sweetened condensed milk and chocolate chips. Heat, stirring often, until the mixture is completely melted and smooth. Make sure to keep the heat low, as the sauce burns easily. Transfer the chocolate sauce to a serving dish. Place it on the table alongside the fruit bouquet and encourage guests to dip their fruit pieces into the sauce.

Try these variations:

- Try placing more than one kind of fruit on a skewer. For example, alternate green and purple grapes on a skewer, or use banana rounds and Clementine segments on a skewer.
- Create a bouquet with one kind of fruit, or make a berries-only arrangement using raspberry and blueberry spears. Spears could also be made out of small strawberries.
- For a fun Fourth of July display, make a red, white and blue bouquet using raspberry, banana and blueberry spears. Arrange them in a holiday-themed canister.

Melon Mania

You will need:

- 1 small to medium watermelon
- 1 to 2 cantaloupes
- 1 to 2 honeydew melons
- Flower-shaped metal cookie cutters
- Melon baller
- Ripple potato slicer
- 15 to 20 (10″) bamboo skewers
- 1 head iceberg lettuce
- 1 small bunch purple or green kale or leafy lettuce
- 1 medium wide-mouth vase or urn

To Begin...

1 Begin by slicing the watermelon, cantaloupe and honeydew melons sideways into ¾"- to 1"-disks. Cut the melons over a sheet pan with a rimmed edge to catch the juice. For a medium-size bouquet, you will need 16 to 20 various melon disks. To cut the flowers, center one of the flower-shaped cookie cutters over a melon disk. (Metal cookie cutters are recommended for a clean, even cut.) Press straight down on the cookie cutter, using even pressure.

2 Cut small-, medium- and large-size flowers out of each type of melon. Turn the melon disks over and gently press the flower shapes out of the disks. Place the melon flowers in an even layer on a clean rimmed baking sheet; place in the refrigerator to chill while assembling the remaining pieces.

3 Use the remaining pieces of watermelon, cantaloupe and honeydew melon to make the flower centers. Cut balls from the melon flesh using a melon baller. The balls can be either completely round or they can have one slightly flat side. For a medium-size bouquet, you will need 8 to 10 flower centers. Place the melon balls on a plate and refrigerate to chill while assembling the melon leaves.

4 Next, cut the melon leaves. You will need half of a cantaloupe and half of a honeydew melon to make enough leaves for a medium-size bouquet. Cut the melons into wedges that measure about 1″ to 1½″ on the widest side. If a rippled effect is desired, cut the melon wedges using a ripple potato slicer, as shown. Run a knife as close to the outer edge of the melon flesh as possible in order to remove the rind. Refrigerate the melon leaves.

5 Next, prepare the lettuce base. Cut the head of lettuce as necessary to fit easily into the vase. Place the lettuce head in the vase so the top of the lettuce sits 1″ to 2″ above the rim of the vase. Stick purple or green kale leaves into the basket to cover the lettuce. The skewers of fruit added later will hold the kale in place.

6 Stack various sizes and colors of melon flowers to create multiple looks. Pierce the center of one large base flower, pressing from the bottom side, through the center, to the top side of the flower. Slide a smaller flower down onto the skewer so it sits on top of the base flower. If desired, top with another smaller flower. If necessary, use rubber bands, gumdrops or raisins to keep the flowers from sliding down the skewers, as explained on page 5.

7 Continue sliding the flowers down the skewer until about ½″ of the skewer is exposed on the top side. Press one melon ball onto the exposed end of the skewer. Then, arrange the melon flowers in the vase. Pierce the skewers through the kale and into the lettuce until the flowers sit at desired height. Sometimes, it is easier to place the melon flowers in the bouquet by first sticking skewers into the kale and then sliding the melon flowers and melon balls onto the skewers.

8 As the vase is filling with flowers, carefully slide the melon leaves onto skewers. Stick the skewers into the arrangement around the rim of the vase. When you are happy with your bouquet, carefully package it for delivery or return the entire arrangement to the refrigerator until ready to display and serve.

Try these variations:

- Try stacking all similar melon pieces together to create single-colored flowers.
- For a true all-melon display, use a hollowed-out watermelon for the base, as shown on page 50.
- If available, try incorporating other fruits from the melon family into the arrangement, such as Canary melons, Christmas melons, Crenshaw melons, Galia melons or Persian melons.

You will need:

- 6 to 10 various-colored apples
- Ripple potato slicer
- 20 to 30 (10″) bamboo skewers
- 1 head iceberg lettuce
- 1 small bunch purple or green kale or leafy lettuce
- 1 medium wide-mouth vase or urn

To Begin...

1 Begin by slicing the apples into wedges. Choose apples of various color and sweetness. For the display shown, Red Delicious, Golden Delicious, Granny Smith and Gala apples were used. Other appropriate apples to use are: Ambrosia, Braeburn, Criterion, Ginger Gold, Honeycrisp and Pink Lady.

2 If a rippled effect is desired, cut the apple wedges using a ripple potato slicer, as shown above. Cut each apple into approximately 6 to 8 wedges, depending on your desired thickness. For a medium-size bouquet, you will need 50 to 60 apple wedges. Once all the wedges are cut, use a paring knife to cut away any of the core, seeds or stem still attached to the apples.

3 To keep the apples from browning, place them immediately on a plate and sprinkle them with lemon juice. Allow the wedges to sit in the juice for about 30 seconds, then turn each wedge over and sprinkle them again. After 1 minute, transfer the apples to a clean plate and set aside.

4 Break each skewer in half so you have approximately 40 to 60 (5″) skewers. Short skewers are desirable since the apple wedges sit close to the base. Insert one short skewer into one pointed end of each apple wedge. Press the skewers 1½″ to 2″ vertically into the apple wedges. Place the skewered apples back on the plate and set aside while assembling the lettuce base.

5 Next, prepare the lettuce base. Cut the head of lettuce as necessary to fit easily into the vase. Place the lettuce head in the vase so the top of the lettuce sits 1″ to 2″ above the rim of the vase. Stick purple or green kale leaves into the vase to cover the lettuce. The skewers of fruit added later will hold the kale in place.

6 Finally, arrange the apple skewers in the vase, alternating types and colors of apples. Stick the skewers into the arrangement in rows around the rim of the vase. Continue filling the bouquet, turning the vase often while arranging to make sure it is filled evenly on all sides. When you are happy with your bouquet, carefully package it for delivery or return the entire arrangement to the refrigerator until ready to display and serve.

Caramel Dipping Sauce

The only thing that could make crisp, juicy apples even sweeter is this decadent caramel dipping sauce. Serve or give it with your Autumn Apples bouquet for a crowd-pleasing presentation. To soften the caramel sauce, simply microwave it for 20 seconds and stir.

Ingredients

½ C. butter
1 (14 oz.) bag caramels, unwrapped
1 (14 oz.) can sweetened condensed milk

Directions

In a medium saucepan over low heat, combine the butter, caramels and sweetened condensed milk. Heat, stirring often, until the mixture is completely melted and smooth. Make sure to keep the heat low, as the sauce burns easily. Transfer the caramel sauce to a serving dish. Place it on the table alongside the fruit bouquet and encourage guests to dip their apple wedges into the sauce.

Try these variations:

- To match a certain color scheme, fill the bouquet with just one type of apple, or create a green and red holiday display using green and red apples.
- Include pear wedges in your bouquet. Anjou and Asian pears work best.
- Serve this bouquet with various dipping sauces, such as peanut butter caramel sauce or the chocolate dipping sauce shown on page 17.

Catching Snowflakes

You will need:

- 17 to 20 strawberries
- 2 to 3 kiwis
- 1 pineapple
- 1 small bunch purple grapes
- Miniature marshmallows
- Snowflake-shaped metal cookie cutter
- 35 to 40 (10″) bamboo skewers
- 1 head iceberg lettuce
- 1 small bunch purple or green kale or leafy lettuce
- 1 (6″ tall) decorative metal bucket

To Begin...

1 Begin by slicing the pineapple sideways into ¾"- to 1"-disks. Cut the pineapple over a sheet pan with a rimmed edge to catch the juice. For a medium- size bouquet, you will need 4 to 6 pineapple disks. To cut the snowflakes, center the snowflake-shaped cookie cutter over a pineapple disk. (Metal cookie cutters are recommended for a clean, even cut.) Press straight down on the cookie cutter, using even pressure.

2 Turn the pineapple disk over and gently press the snowflake shape out of the disk. Cut the remaining disks into snowflakes and place them in an even layer on a clean rimmed baking sheet. Insert a skewer into the crevice between two points of the pineapple snowflake. Slide the snowflake vertically down onto the skewer until the skewer hits the hard center of the pineapple piece. Refrigerate the snowflakes while assembling the remaining pieces. Any remaining pieces of the cut pineapple disks can be discarded.

3 To make the grape-mallow spears, alternately thread grapes and miniature marshmallows onto a wooden skewer, starting and ending with a grape. Do not pierce all the way through the final grape on each spear, allowing the skewer to remain concealed. For a medium-size bouquet, you will need about 6 grape-mallow spears. Place the spears on a plate and refrigerate to chill.

4 Cut each kiwi in half to make the kiwi flowers. Using a paring knife, gently peel and discard the outer brown skin from each kiwi half. Stick a skewer into the thickest part of each kiwi half, pressing into the rounded side and stopping before the skewer pierces through the flat side of the kiwi. If necessary, use rubber bands, gumdrops or raisins to keep the flowers from sliding down the skewers, as explained on page 5. Place the kiwi flowers on a plate and refrigerate.

5 To make the strawberry-mallow blossoms, rinse the strawberries under cool water and pat dry gently with paper towels. Choose large, full strawberries of a similar size. If desired, remove the stem and leaves from each strawberry, however it is not necessary to do so. Cut a slit ⅔ of the way into each strawberry, starting at the pointed end. Stick a skewer into each berry from the stem end. Slide 1 or 2 miniature marshmallows onto the exposed end of each skewer.

6 Next, prepare the lettuce base. Cut the head of lettuce as necessary to fit easily into the bucket. Place the lettuce head in the bucket so the top of the lettuce sits 1″ to 2″ above the rim of the bucket. Stick purple or green kale leaves into the bucket to cover the lettuce. The skewers of fruit added later will hold the kale in place.

7 Remove all the fruit pieces from the refrigerator and start arranging the fruit flowers in the bucket. Place the pineapple snowflakes first, followed by the strawberry-mallow blossoms. Once most of the strawberries are in place, arrange the kiwi flowers in the bucket, flat side facing up.

8 As the basket is filling with flowers, carefully stick the grape-mallow spears into the arrangement to fill any gaps. Turn the bucket often while arranging to make sure it is filled evenly on all sides. When you are happy with your bouquet, carefully package it for delivery or return the entire arrangement to the refrigerator until ready to display and serve.

Try these variations:

- For a bright and cheery bouquet, use pastel-colored miniature marshmallows.
- Include plain strawberry buds in the bouquet, as well as strawberry-mallow blossoms. The technique for making strawberry buds is described on page 11.
- If you prefer a fruit-only bouquet, replace the miniature marshmallows with small slices of banana. Use them to make the strawberry-mallow blossoms and grape-mallow spears.

Shooting Star

You will need:

- 1 whole pineapple (just 1 disk needed)
- 1 pint blueberries
- 8 to 10 strawberries
- ½ honeydew melon
- ½ cantaloupe
- Star-shaped metal cookie cutter

- 25 to 30 (10″) bamboo skewers
- 1 head iceberg lettuce
- 1 small bunch purple or green kale or leafy lettuce
- 1 medium vase, ceramic planter or urn

To Begin...

1 Begin by cutting one ¾"- to 1"-disk from the pineapple. Use the remaining pineapple for other purposes. To cut the star, center the star-shaped cookie cutter over the pineapple disk. (Metal cookie cutters are recommended for a clean, even cut.) Press straight down on the cookie cutter, using even pressure.

2 Turn the pineapple disk over and gently press the star shape out of the disk. Any remaining pieces of the cut pineapple disks can be discarded. Insert a skewer into the crevice between two points of the pineapple star. Slide the star vertically down onto the skewer until the skewer hits the hard center of the pineapple piece. Set the pineapple star on a clean plate; place in the refrigerator to chill while assembling the remaining pieces.

3 Next, make the blueberry spears. Thread 6 to 8 similar-size blueberries onto a wooden skewer, starting at the bottom-side of each blueberry and piercing straight through to the top end of each berry. Do not pierce all the way through the final berry on each spear, allowing the skewer to remain concealed. For a medium-size bouquet, you will need 8 to 10 blueberry spears. Place the berry spears on a plate and refrigerate to chill while assembling the base.

4 To create the strawberry buds, rinse the strawberries under cool water and pat dry gently with paper towels. Choose large, full strawberries of a similar size. If desired, remove the stem and leaves from each strawberry, however it is not necessary to do so. Poke the pointed end of a skewer into the stem end of a strawberry, stopping before the skewer pierces through the berry; repeat with remaining berries and skewers, then refrigerate.

5 Next, cut the melon leaves. You will need half of a cantaloupe and half of a honeydew melon to make enough leaves for a medium-size bouquet. Cut the melons into wedges that measure about 1″ to 1½″ on the widest side. If a rippled effect is desired, cut the melon wedges using a ripple potato slicer. Run a knife as close to the outer edge of the melon flesh as possible in order to remove the rind. Finally, cut each wedge in half to make two leaves. Refrigerate the melon leaves.

6 Prepare the lettuce base. Cut the head of lettuce as necessary to fit easily into the urn. Place the lettuce head in the urn so the top of the lettuce sits 1″ to 2″ above the rim of the urn. Stick purple or green kale leaves into the urn to cover the lettuce. The skewers of fruit added later will hold the kale in place.

7 Begin arranging the fruit pieces in the urn. Carefully slide the melon leaves onto skewers. Place a row of melon leaves around the rim of the urn, curved side facing down, alternating between cantaloupe and honeydew melon leaves. Stick the strawberry buds into the urn in a circle just inside the row of melon leaves.

8 Slide about 6 blueberries onto the skewer holding the pineapple star in order to conceal the skewer. Stick the pineapple star into the center of the bouquet. Surround the star with desired amount of blueberry spears. When you are happy with your bouquet, carefully package it for delivery or return the entire arrangement to the refrigerator until ready to display and serve.

Try these variations:

- For a larger bouquet, add several pineapple stars in various sizes.
- Include other spears in the arrangement, such as raspberry spears or grape spears.
- Create a Fourth of July bouquet by eliminating the melon leaves and arranging the remaining fruit in a smaller vase. Garnish the arrangement with small American flags.

Citrus
Citrus Smiles

You will need:

- 1 pear
- 1 small bunch purple globe grapes
- 3 to 4 Clementines
- 1 to 2 large oranges
- 6 to 8 strawberries
- 20 to 30 (10″) bamboo skewers
- 1 head iceberg lettuce
- 1 small bunch purple or green kale or leafy lettuce
- 1 medium ceramic canister or jar

To Begin...

1 Begin by creating the strawberry buds. Rinse the strawberries under cool water and pat dry gently with paper towels. Choose large, full strawberries of a similar size. If desired, remove the stem and leaves from each strawberry, however it is not necessary to do so. Poke the pointed end of a skewer into the stem end of a strawberry stopping before the skewer pierces through the berry; repeat with remaining berries and skewers, then refrigerate.

2 Peel the oranges and Clementines, then use a paring knife to remove as much of the white pith as possible. Gently separate the oranges and Clementines into halves. Set aside one of the orange halves and two of the Clementine halves. Separate the remaining oranges and Clementines into individual segments.

3 Next, create several styles of orange and Clementine spears. Make sailboat spears by sliding two Clementine segments sideways onto a skewer so the segments form a boat shape. Create several tall spears by sliding orange segments vertically onto a skewer. Finally, make a few Clementine-grape spears by alternately sliding Clementine segments and grapes onto a skewer.

4 Use the reserved orange and Clementine halves to create the citrus blossoms. Set a single grape in the indentation of one orange half, pressing down lightly on the grape. Press a skewer into the orange from the bottom, through the grape, then back into the orange flesh so the flat side of the orange is facing forward and the grape is secured into the center of the orange. Repeat with the remaining Clementine halves. For a medium-size bouquet, you will need 3 to 5 citrus blossoms.

5 Gently rinse the pear under cool water and pat dry with paper towels. Cut the pear into 6 even segments. Use a paring knife to cut a shallow curve into the pear removing the core area and any seeds. Set a single grape in the indentation of one pear segment. Holding the grape in place, stick a skewer into the pear segment from the bottom end through the grape and back into the pear flesh so the pear is sitting vertically and the grape is secured in the center groove of the pear.

6 Next, prepare the lettuce base. Cut the head of lettuce as necessary to fit easily into the canister. Place the lettuce head in the canister so the top of the lettuce sits 1″ to 2″ above the rim of the canister. Stick purple or green kale leaves into the canister to cover the lettuce. The skewers of fruit added later will hold the kale in place.

7 Start arranging the fruit by placing several sailboat spears around the rim of the canister. Place a row of strawberry buds behind the sailboat spears. Arrange the pear segments around the bouquet, turning the canister often while arranging to make sure it is filled evenly on all sides. Stick the orange and Clementine halves into the arrangement, as well as the Clementine-grape spears.

8 If a playful look is desired, the pieces can be arranged to resemble a smiley face, with the Clementine halves placed as eyes and the sailboat spears placed as a mouth, similar to the arrangement on page 34. When you are happy with your bouquet, carefully package it for delivery or return the entire arrangement to the refrigerator until ready to display and serve.

Try these variations:

- Place 3 or 4 Clementine segments on a skewer to make taller sailboat spears. Instead of placing them around the rim of the canister, place them closer to the center.
- Create a classy-looking display by lining the rim of the canister with several Clementine halves.
- Incorporate green grapes for more color.
- Use grapefruit segments in the display. Serve chocolate dipping sauce (recipe on page 17) on the side to sweeten up the taste of the grapefruit.

Veggie Delight

You will need:

- 3 to 4 green, yellow and red bell peppers
- 1 small bunch cauliflower
- 1 small bunch broccoli
- 1 cucumber
- 1 zucchini
- 5 to 7 large carrots
- 5 to 7 large radishes
- 10 to 12 cherry tomatoes
- 1 bunch green onions

- Ripple potato slicer
- Vegetable coring tool
- Small flower-shaped cookie cutter
- 45 to 60 (10″) bamboo skewers
- 1 head iceberg lettuce
- 1 small bunch purple or green kale or leafy lettuce
- 1 medium basket

To Begin...

1 Begin by making the carrot spears. Peel and wash each carrot, then cut off both ends of the carrot to make two flat ends. Using the ripple potato slicer, cut each carrot through the center at a diagonal, creating two carrot pieces. Stick a skewer into the flat end of each carrot piece. Separate the cleaned broccoli and cauliflower into small florets. Stick a skewer into the bottom of each floret; set aside the vegetable pieces in the refrigerator.

2 To make the cucumber flowers, slice the cleaned and dried cucumber into ½"-thick rounds. To cut a flower, center the flower-shaped cookie cutter over a cucumber round. Press straight down on the cookie cutter, using even pressure. Stick a skewer into a cucumber flower, pressing from the flat bottom side, through the center, to the top side of the flower. Slide one cherry tomato onto the exposed end of the skewer.

3 Next, create the bell pepper flowers. Wash and pat dry 1 or 2 green bell peppers. Cut the peppers in a zig-zag pattern horizontally around the center. Carefully pull the two sides apart and remove the inner membrane and seeds. Use a few of the cucumber flowers as centers for the bell pepper flowers. Stick a skewer through the bottom of the pepper, through a cucumber flower, then into a cherry tomato.

4 Use the coring tool to cut a hole entirely through a radish. Remove the core, leaving the outer radish as a tube. With the coring tool still in the radish, cut vertical lines around the radish tube, stopping about halfway down the side. Stick a cherry tomato snuggly inside each radish tube. Press the radish flower onto a skewer by sticking it into the tomato from the bottom side.

5 Wash and pat dry the zucchini. To make the zucchini flowers, cut the zucchini into 2″ segments. Run a knife vertically between the outer zucchini peel and inner flesh to create the petals, stopping about 1½″ down the side of the zucchini piece; repeat to make 5 petals. Carefully remove the inner flesh in the center of the petals, leaving a ½″ base of zucchini intact. Stick a skewer up through the bottom of the flower and top with a cherry tomato.

6 Cut the yellow and red bell peppers into thin wedges that are the length of the pepper and about ¾″ wide in the center. Carefully slide the peppers onto skewers, piercing once at the pointed end of a pepper wedge and poking gently into the other pointed end.

7 Make thin green onion leaves by running a knife through each onion to make two halves; cut the ends into points. Next, prepare the lettuce and kale base as described in previous arrangements.

8 Finally, fill the bouquet with the vegetable flowers and leaves. When you are happy with your bouquet, carefully package it for delivery or return the entire arrangement to the refrigerator until ready to display and serve.

Creamy Dill Dip

Serve or give your Veggie Delight bouquet with this delicious dill dip. Those vegetables will disappear in the blink of an eye!

Ingredients

4 oz. reduced-fat cream cheese, softened
½ pkg. dry ranch dressing mix
2 T. skim milk
1½ tsp. dried dillweed or 1 T. chopped fresh dill

Directions

In a food processor, combine the cream cheese, ranch dressing mix, milk and dillweed. Process on medium speed until blended and smooth. Store the dip, tightly covered, in the refrigerator until ready to serve. Transfer the dill dip to a serving dish. Place on the table alongside the bouquet and encourage guests to dip their veggies.

Regal Relishes

You will need:

- **20 to 25 cherry tomatoes**
- **1 each yellow and red bell peppers**
- **5 to 7 long celery ribs**
- **1 large bunch broccoli**
- **1 (16 oz.) bag rippled carrot chips**
- **5 to 7 radishes**
- **25 to 30 (10″) bamboo skewers**
- **1 head iceberg lettuce**
- **1 small bunch purple or green kale or leafy lettuce**
- **1 small to medium oval basket**

To Begin...

1 Begin by breaking each skewer in half so you have approximately 50 to 60 (5″) skewers. Short skewers are desirable since the vegetables sit close to the base. Wash and pat dry the tomatoes. Stick one short skewer gently into the stem end of each cherry tomato, stopping before the skewer pierces through the tomato. Set the tomato buds on a plate and chill in the refrigerator while preparing the remaining pieces.

2 After washing and patting dry the yellow and red bell peppers, cut them into wedges that are the length of the pepper and about ¾″ wide in the center. Carefully slide the peppers onto skewers, piercing once at the pointed end of a pepper wedge and poking gently into the other pointed end; refrigerate.

3 Rinse and pat dry the celery ribs. Cut both ends of a celery rib into a point, then cut the rib in half in the middle creating two pointed celery pieces. Repeat with the remaining celery ribs. Gently slide a skewer into the flat end of each celery point, stopping before the skewer pierces through the celery. Refrigerate the celery points.

4 Separate the cleaned broccoli into small florets. Stick a skewer into the bottom of each floret; set aside in the refrigerator. To make the carrot spears, choose 10 to 16 long and thick carrot chips from the bag. Gently slide the pointed end of a small skewer into the thickest end of a carrot chip; set aside in the refrigerator.

5 To make the radish rosettes, wash and gently pat dry the radishes. Cut off both ends of each radish to make two flat sides with the white inner radish exposed. Use a paring knife to cut ½" petals along the side of each radish. Place the radish rosettes in a bowl of ice water to help open up the petals, as shown below.

6 Next, prepare the lettuce base. Cut the head of lettuce as necessary to fit easily into the basket. For a small to medium oval-shaped basket, cut about 2" from both sides of the lettuce head. Use the cut-off pieces to fill the bottom and sides of the basket, then place the lettuce head in the basket so the top of the lettuce sits 1" to 2" above the rim of the basket. Stick purple or green kale leaves into the basket to cover the lettuce.

7 Arrange the vegetable skewers in sections. Create two rows of tomato buds along the basket rim on one long side. Stick the celery and carrot spears in rows around both short sides of the basket. Create two rows of pepper wedges along the other long side of the basket. Fill in the center with broccoli florets and radish rosettes. When you are happy with your bouquet, carefully package it for delivery or return the entire arrangement to the refrigerator until ready to display and serve.

Ranch Veggie Dip

The crowning detail for your Regal Relish bouquet is this tasty ranch veggie dip. The dip can be made ahead of time and stored in the refrigerator for up to 10 days.

Ingredients
1 C. reduced-fat mayonnaise
½ C. reduced-fat sour cream

½ tsp. dried chives
½ tsp. dried parsley flakes
½ tsp. dried dillweed
¼ tsp. ground garlic
¼ tsp. onion powder
⅛ tsp. salt
⅛ tsp. pepper

Directions
In a bowl, mix together the mayonnaise, sour cream, chives, parsley, dillweed, ground garlic, onion powder, salt and pepper; mix until blended and smooth. Store the dip, tightly covered, in the refrigerator until ready to serve. Transfer the ranch dip to a serving dish. Place it on the table alongside the bouquet and encourage guests to dip their veggies.

You will need:

- 35 to 40 strawberries
- 1 (16 oz.) pkg. microwaveable chocolate candy coating
- 1 (16 oz.) pkg. microwaveable vanilla candy coating
- Pink gel food coloring
- 12″ disposable decorating bags with optional small tip (#3)

- Rainbow sprinkles or nonpareils
- 35 to 40 (10″) bamboo skewers
- 1 head iceberg lettuce
- 1 small bunch purple or green kale or leafy lettuce
- 1 clay pot
- Decorative ribbon and bows

To Begin...

1 Begin by creating the strawberry buds. Rinse the strawberries under cool water and pat dry gently with paper towels. Choose large, full strawberries of a similar size. If desired, remove the stem and leaves from each strawberry, however it is not necessary to do so. Poke the pointed end of a skewer into the stem end of a strawberry stopping before the skewer pierces through the berry; repeat with remaining berries and skewers, then refrigerate.

2 Follow the package directions to melt the chocolate candy coating in the microwave. Depending on the size of your bouquet, melt the entire amount of candy coating or just portions of it. Remove the strawberries from the refrigerator and pat them dry with paper towels. It is preferable to dip cold, dry strawberries. Holding a strawberry bud by the skewer, dip it into the melted chocolate and use a spoon to help coat the strawberry completely.

3 Hold the dipped strawberry over the melted chocolate, allowing any excess to drip off. In order for the chocolate to dry around the strawberries with a smooth finish on all sides, stick the non-berry end of each skewer into a sheet of Styrofoam. To create the sprinkled strawberries, hold a fresh dipped strawberry over a bowl and sprinkle the nonpareils lightly over the coating. Set the coated berries in the Styrofoam to dry completely.

4 Follow the package directions to melt the vanilla candy coating in the microwave. Holding a strawberry bud by the skewer, dip it into the white coating and use a spoon to help coat the strawberry completely. Create 4 to 6 vanilla-coated berries. Stick the skewers into the Styrofoam, allowing the coating to dry.

5 Using a toothpick, add a few drops of the pink gel food coloring to the melted vanilla coating. Quickly stir the coloring into the coating until it turns a light shade of pink. If necessary, add more coloring until your desired shade is achieved. Dip some of the strawberry buds into the pink coating, using a spoon to help coat each strawberry completely.

6 Working quickly, transfer spoonfuls of the melted pink coating (white coating can also be used) into a plastic decorating bag. If using a tip, fit the tip and/or coupler into the bag before filling it with the coating. Snip a small hole into the tip of the bag with a scissors. To make the dotted berries, hold the skewer end of one chocolate-dipped berry in one hand and the filled decorating bag in the other hand. Dot the pink coating all around the strawberry. Create the swirled berries in the same fashion, moving the pink coating in a swirling pattern around the berry.

7 Next, prepare the lettuce base. Cut the head of lettuce as necessary to fit easily into the pot. Place the lettuce head in the pot so the top of the lettuce sits 1″ to 2″ above the rim of the pot. Stick purple or green kale leaves into the pot to cover the lettuce. The skewers of fruit added later will hold the kale in place. If desired, wrap a decorative ribbon around the pot.

8 Once all the coating has dried, arrange the berries in the pot. Stick the berries very close together to achieve a full, balanced bouquet. Alternate colors and styles of decorative berries while arranging, turning the pot often to make sure it is filled evenly on all sides. If desired, decorate the bouquet or skewers with small bows made from ribbon. When you are happy with your bouquet, carefully package it for delivery or return the entire arrangement to the refrigerator until ready to display and serve.

Try these variations:

- Use various colors and shades of gel food coloring to make dipped strawberries in any color. Make a bouquet to match any holiday or event.
- For an amusing Super Bowl or tailgate party bouquet, create Berry Footballs using the technique shown on page 60. Add the Berry Footballs to your arrangement along with strawberries dipped in team colors.
- Create a sweet baby shower bouquet with pink and light blue dipped strawberries. Write the word "Girl" on the pink berries and "Boy" on the blue berries. Each guest can choose a berry revealing their prediction for the expecting mother.

Feelin' Fruity

You will need:

- 1 personal-size small watermelon
- 1 whole pineapple
- 1 large orange
- 10 to 12 strawberries
- 1 small bunch purple grapes
- ½ cantaloupe
- ½ honeydew melon
- Flower-shaped cookie cutters and melon baller

- 1 (16 oz.) pkg. microwaveable chocolate candy coating
- 12″ disposable decorating bag with optional small tip (#3)
- 25 to 35 (10″) bamboo skewers
- 1 head iceberg lettuce
- 1 small bunch purple or green kale or leafy lettuce

To Begin...

1 Begin by cutting a thin slice off the bottom of the watermelon so it rests flat and securely on the table. Cut an oval shape out of the top of the watermelon to create an opening. Use a spoon to remove all the watermelon flesh and juice to a bowl (see page 53 for a refreshing watermelon smoothie recipe). Fit the head of lettuce into the hollow watermelon shell, cutting the lettuce as necessary. Stick purple or green kale leaves into the watermelon to cover the lettuce; set aside.

2 Cut the pineapple flowers and cantaloupe flower centers as described on page 7. Slice the unpeeled orange into about 6 wedges. Stick a skewer vertically into each orange wedge. Create the strawberry buds as described on page 11. Place the pineapple flowers, cantaloupe centers, orange wedges and strawberry buds on a large plate; chill in the refrigerator while preparing the chocolate coating.

3 Follow the package directions to melt the chocolate candy coating in the microwave. Depending on the size of your bouquet, melt the entire amount of candy coating or just portions of it. Dip half of a pineapple flower in the chocolate coating; repeat as desired. Dip a few of the strawberry buds in chocolate as described on page 47. Fill the decorating bag with some of the melted chocolate, snip a hole in the end of the bag and decorate the orange wedges with chocolate, moving in a swirling pattern.

4 To make the grape spears, thread 4 or 5 similar-size grapes onto a wooden skewer, starting at the stem-side of each grape and piercing straight through to the bottom end of each grape. Do not pierce all the way through the final grape on each spear, allowing the skewer to remain concealed. For a small-size bouquet, you will need 4 to 6 grape spears. Place the grape spears on a plate and refrigerate to chill.

5 Next, make the melon leaves. Cut the melons into wedges that measure about 1″ to 1½″ on the widest side. If a rippled effect is desired, cut the melon wedges using a ripple potato slicer. Run a knife as close to the outer edge of the melon flesh as possible in order to remove the rind. Insert a skewer vertically into each melon leaf. Refrigerate the melon leaves.

6 Assemble the pineapple daisies as described on page 8, using both plain and chocolate-dipped pineapple flowers. Stick the pineapple daisies into the lettuce-filled watermelon. Arrange the remaining fruit pieces in the bouquet, turning the watermelon often to make sure it is filled evenly on all sides. When you are happy with your bouquet, carefully package it for delivery or return the entire arrangement to the refrigerator until ready to display and serve.

Watermelon Smoothie

*Don't let all that watermelon scooped out of the base of
your Feelin' Fruity bouquet go to waste. Blend it into a
refreshing smoothie for a cool summer treat, or make a big
batch to serve alongside your fruit bouquet!*

Ingredients
2 C. seedless watermelon
   chunks
1 C. cracked or shaved ice
½ C. plain yogurt
1 to 2 T. sugar
½ tsp. ground ginger
⅛ tsp. almond extract

Directions
In a blender, combine the watermelon chunks, ice,
yogurt, sugar, ground ginger and almond extract. Process
on medium speed until blended and smooth. Pour into 2
to 3 glasses. If desired, garnish the rim of each glass with
small strawberries or a watermelon wedge.

Try these variations:

- Use a medium- to large-size watermelon for the base.
  Fill it with fruit to feed a crowd.

- Make the base out of a hollowed cantaloupe, honeydew
  melon or pineapple. If using a pineapple, keep the
  crown intact and lay the pineapple on one side, cutting
  as necessary for the pineapple to sit flat on the table.

- This bouquet would make a wonderful addition to a
  summer picnic or potluck. Assemble the bouquet at
  home, omitting any of the chocolate-dipped pieces,
  and carefully transport it to the event in a large cooler.
  Set the bouquet inside a small clear plastic or glass
  bowl. Fill the bowl with a little ice to keep the fruit
  cool. Encourage picnickers to enjoy the fruit within
  30 minutes.

Straw-Kiwi Craze

You will need:

- 4 to 5 kiwis
- 20 to 25 strawberries
- 1 (16 oz.) pkg. microwaveable chocolate candy coating
- 12″ disposable decorating bag with optional small tip (#3)
- 20 to 30 (10″) bamboo skewers
- 1 head iceberg lettuce
- 1 small bunch purple or green kale or leafy lettuce
- 1 (5½″ tall) square container or vase

To Begin...

1 Begin by creating the strawberry buds. Rinse the strawberries under cool water and pat dry gently with paper towels. Choose large, full strawberries of a similar size. If desired, remove the stem and leaves from each strawberry, however it is not necessary to do so. Poke the pointed end of a skewer into the stem end of a strawberry stopping before the skewer pierces through the berry; repeat with remaining berries and skewers, then refrigerate.

2 Follow the package directions to melt the chocolate candy coating in the microwave. Depending on the size of your bouquet, melt the entire amount of candy coating or just portions of it. Remove the strawberries from the refrigerator and pat them dry with paper towels. It is preferable to dip cold, dry strawberries. Holding a strawberry bud by the skewer, dip it into the melted chocolate and use a spoon to help coat the strawberry completely.

3 Dip some of the berries entirely in chocolate, but leave most of the berries un-dipped. Working quickly, transfer spoonfuls of the melted chocolate into a plastic decorating bag. If using a tip, fit the tip and/or coupler into the bag before filling. Snip a small hole into the tip of the bag with a scissors. Hold the skewer end of one un-dipped berry in one hand and the filled decorating bag in the other hand. Move the chocolate in a swirling pattern around the berry. Stick the skewers into a sheet of Styrofoam to allow the chocolate to harden.

4 Cut each kiwi in half to make the kiwi flowers. Using a paring knife, gently peel and discard the outer brown skin from each kiwi half. Stick a skewer vertically into one end of each kiwi half, pressing through the kiwi and stopping before the skewer pierces through the other end of the kiwi. Place the kiwi flowers on a plate and refrigerate.

5 Next, prepare the lettuce base. Cut the head of lettuce as necessary to fit easily into the container. Place the lettuce head in the container so the top of the lettuce sits 1″ to 2″ above the rim of the container. Stick purple or green kale leaves into the container to cover the lettuce. The skewers of fruit added later will hold the kale in place.

6 Once all the coating has dried, begin arranging the bouquet. Place a row of swirled berries around the inside rim of the container. Place a row of kiwi flowers above the swirled berries, followed by a row of chocolate-dipped berries. Top off the center of the bouquet with one large swirled berry. When you are happy with your bouquet, carefully package it for delivery or return the entire arrangement to the refrigerator until ready to display and serve.

Got Melted Chocolate?

If you have melted chocolate leftover after creating your fruit bouquet, try using it up with one of the following ideas. If the chocolate needs to be re-melted, just heat it in the microwave for 20 seconds and stir; repeat as necessary.

- Pour the chocolate into candy molds to make small treats or favors.
- Using a small pastry brush, paint the melted chocolate onto one side of a clean plastic leaf. Once the chocolate has hardened, peel away the leaf to reveal your textured chocolate leaf; use it to garnish a bowl of ice cream or special dessert.
- Make chocolate cups by painting two to three layers of melted chocolate on the inside of paper cupcake liners or miniature candy liners. Once the chocolate has hardened, peel away the liners. Fill the chocolate cups with pudding, mousse or fresh berries.
- Place a sheet of waxed paper on a flat surface. Using a decorating bag, draw small chocolate letters, numbers and shapes on the waxed paper. Once the chocolate has hardened, carefully lift away the shapes. Or, write the initials of each dinner guest in chocolate and use them to create personalized desserts.
- Dip other foods, such as twisted pretzels, long straight pretzels, maraschino cherries, sandwich cookies, chow mein noodles, graham crackers or marshmallows. Shake off any excess chocolate and set the dipped foods on waxed paper while the chocolate hardens.
- Make chocolate stirring spoons. Dip the head of each spoon into the chocolate and let dry. If desired, sprinkle the chocolate with nonpareils, colored  sugar or miniature marshmallows before the chocolate hardens. Wrap the chocolate end of the spoons with a small piece of plastic wrap and tie off with a ribbon. Give the spoons as gifts for friends to stir into their coffee or hot chocolate.

Chocolate Creations

For an up-scale party, don't just drape your strawberries in chocolate; dress them in classy chocolate tuxedos. Incorporate the strawberry tuxedos into your fruit bouquet, or simply line them in rows on a serving tray for a stunning presentation.

Tuxedos

1 Begin by dipping strawberry buds into melted vanilla candy coating (directions for creating strawberry buds and melting vanilla coating described on pages 47 and 48). In order for the vanilla coating to dry around the strawberries with a smooth finish on all sides, stick the non-berry end of each skewer into a sheet of Styrofoam, as shown on page 47.

2 Once the vanilla coating has hardened, melt the chocolate coating according to package directions. Holding one dipped strawberry by the skewer, dip it sideways into the chocolate coating to create the front of the tuxedo jacket; repeat on other side. The vanilla coating "V" should still be visible as the white dress shirt below the jacket.

3 When all the jackets have been dipped, quickly spoon the melted chocolate into a decorating bag. Snip a small hole in the end of the bag or use a tip and/ or coupler. Holding one strawberry by the skewer, draw three small dots down the front of the jacket as the buttons, then draw a small bowtie with the chocolate above the top button. Repeat with remaining berries; set aside to dry completely. The dipped colors can be applied in reverse order to make strawberries with white tuxedo jackets, as shown in the photo.

Chocolate Creations

The sky's the limit when it comes to decorating your fruit bouquets. Accent them with these delightful strawberry balloons. Appropriate for a kid's party or circus-themed event, these strawberry balloons can be used to adorn your fruit bouquet, or fill an entire container with only strawberry balloons in a rainbow of colors.

Flying High

1 Begin by creating strawberry buds as directed on page 47. Open a package of vanilla candy coating and use a large knife to carefully break the block of coating into 4 to 6 pieces. Melt each piece, one at a time, as directed on the package. Using a toothpick, add a few drops of one gel food coloring to the melted vanilla coating. Quickly stir the coloring into the coating until desired shade is achieved. Dip some of the strawberry buds into the colored coating, using a spoon to help coat each strawberry completely; set aside to dry.

2 Repeat the coloring procedure with other colors, creating strawberry balloons in several hues. In order for the colored coating to dry around the strawberries with a smooth finish on all sides, stick the non-berry end of each skewer into a sheet of Styrofoam, as shown on page 47. When the colored berries have dried completely, tie a ribbon bow around each skewer at the base of the strawberry balloon in a coordinating color.

3 For a balloon-only display, fit a piece of Styrofoam into the display container. Cover the Styrofoam with sky-colored tissue paper, then arrange the strawberry balloons in the container. If desired, arrange cotton balls or cotton candy around the base of the balloons to look like clouds.

Chocolate Creations

Score a touchdown with your fruit bouquet by incorporating these berry footballs into the arrangement. Win over your friends and family at the next Super Bowl or tailgate party with a centerpiece full of these chocolate-covered strawberries and other fruits dipped in team colors.

Berry Footballs

1 Begin by dipping strawberry buds into melted chocolate candy coating (directions for creating strawberry buds and melting chocolate described on page 47). In order for the chocolate coating to dry around the strawberries with a smooth finish on all sides, stick the non-berry end of each skewer into a sheet of Styrofoam, as shown on page 47.

---

2 Once the chocolate coating has hardened, melt a small amount of vanilla candy coating according to package directions. The vanilla coating can be applied to the footballs with a toothpick or decorating bag fitted with a tip and/or coupler. Holding one dipped strawberry by the skewer, draw a thin vertical line over the chocolate. Next, draw three cross-lines over the vertical line to represent the stitching on the football. Repeat with remaining berries; set aside to dry completely.

---

3 This same technique can be used to create berry basketballs, baseballs or tennis balls. Dye vanilla coating in the appropriate base color and set aside to dry. Next, dye a small amount of melted vanilla candy coating in an appropriate color for the ball stitching; draw the stitching on the ball using a toothpick or decorating bag. See diagrams below for appropriate stitching patterns.